Invisible Pull: Electricity and Magnets

Harcourt

SCHOOL PUBLISHERS

Orlando   Austin   New York   San Diego   Toronto   London

Visit *The Learning Site!*

www.harcourtschool.com

Static Electricity

Every day, you turn on lights in your home. In winter, you may turn on the heat. You depend on a refrigerator to keep your food cold. Maybe you heat food in a microwave oven. All of these tasks are possible because of electricity. **Electricity** is a form of energy produced by moving electrons.

Static electricity is a kind of electricity that forms when an electric charge builds up in an object. It can even build up in a person. It can make your hair stand straight up.

Without electricity, it would be hard to light your house.

When electricity builds up in a cloud, it comes out as lightning.

Lightning is also a form of static electricity. It occurs when electricity builds up in a cloud. The electricity comes out of the cloud as a bolt of lightning. The lightning releases so much energy that you can hear thunder even from far away.

MAIN IDEA AND DETAILS

Focus Skill

What might make your hair stand straight up?

Current Electricity

Electricity that moves through a wire is called current electricity. When you plug in a lamp, you are connecting the lamp to a wire hidden in the wall. That wire carries current electricity.

A **circuit** is a path that electricity flows through. The bulb in a lamp lights up when there is a complete circuit for the electric current to follow from the wall to the bulb.

The TV and DVD player depend on current electricity to make them work.

Electricity is even used outdoors.

Many appliances in your home depend on electricity. You use current electricity to power objects that are plugged in. Suppose you plug in a TV. Current electricity moves from wires in the wall to the TV's plug and then into the TV itself. The current electricity powers the TV.

Items that are not in the home use current electricity, too. Signs on the outside of buildings, traffic lights, and cars all use current electricity.

 MAIN IDEA AND DETAILS What makes plugged in appliances run?

Conductors and Insulators

There are some things that electricity can move through easily. These things are called *conductors*. For example, electricity moves easily through metal. The wires that current electricity travels through are often made of copper, which is a very good conductor.

This reel contains copper wire, which is often used to carry electricity.

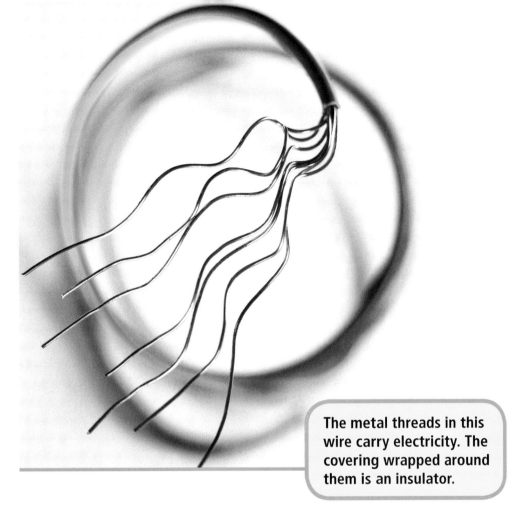

The metal threads in this wire carry electricity. The covering wrapped around them is an insulator.

Other things block electricity and do not allow it to pass through. These things are called *insulators*.

Plastic and rubber make good insulators. That is why they are often used to cover wire. It can be dangerous to touch a wire when it is carrying a current. Insulators prevent you from getting hurt by blocking the electric current.

 COMPARE AND CONTRAST How are conductors and insulators alike? How are they different?

Magnetism

Magnets are made of a kind of metal that attracts iron. Anything that attracts objects with iron in them is **magnetic**.

When a metal is attracted to a magnet, it can stick to the magnet. A magnet can lift an object if the attraction is strong.

Not all metals are attracted to magnets—only those with iron.

These objects are attracted to magnets.

Every magnet has both a
north pole and a south pole.

Every magnet has two ends, or poles. One end
is called the north pole, and the other is called
the south pole. The north pole of one magnet
attracts the south pole of another magnet. They
stick together.

If you try to put the north pole of one magnet
near the north pole of another magnet, the
magnets will push away, or repel, each other.

Opposite poles of magnets are attracted to each
other. Like poles repel each other.

COMPARE AND CONTRAST What happens if you put the
north poles of two magnets near each other? What
happens if you put the north pole of one magnet near the
south pole of another?

Uses of Magnets

Magnets can be used in many ways. Small magnets are used around the house to hold papers on refrigerator doors and to keep cabinets closed. You may have an electric can opener at home that uses a magnet to hold the can in place.

There are many uses for magnets. Some can be fun!

Lodestone is a natural mineral that acts as a magnet. It attracts iron and was once used by sailors to make a simple compass.

A compass contains a magnetic needle that is attracted to Earth's North Pole. A compass can help you figure out which way you need to go.

Magnets are used in a lot of things we need every day. These include computers, motors, and compasses.

MAIN IDEA AND DETAILS Name and describe ways that people use magnets every day.

Combining Magnets and Electricity

Some magnets are powered by electricity. As long as the electricity is on, the magnet attracts things with iron in them. When the electricity is turned off, the magnet stops working. These magnets are called *electromagnets*.

Electromagnets are very helpful when you need to move something very heavy. When the electricity is on, you can lift the object easily. When you turn the electricity off, you can remove the object from the magnet without having to pull.

This ordinary screw can become an electromagnet if wire containing electric current is wrapped around it.

This giant electromagnet lifts large objects.

Electromagnets are very useful to builders. Workers can use an electromagnet to lift a huge, metal beam into place. Then, when the beam is in the right place, they turn off the electricity and the beam will stay put.

Large magnets can also be used to move cars that no longer run and to separate iron from other scrap material for recycling.

 COMPARE AND CONTRAST How is an electromagnet different from a regular magnet?

Electricity from Magnets

Just as electricity can make a magnet, a magnet can make electricity. You can make electricity by moving a magnet near a coil of wire so that current electricity flows through the wire. A machine that makes electricity this way is called a *generator.*

Suppose that power lines near your house get knocked down during a storm. You can't get electricity in your home. If you have your own generator, you'll be able to make your own electricity—from a magnet.

 MAIN IDEA AND DETAILS **How are magnets used to generate electricity?**

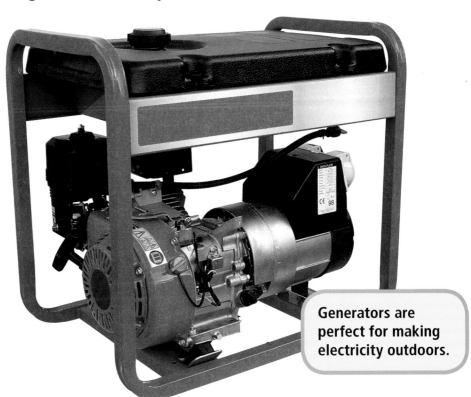

Generators are perfect for making electricity outdoors.

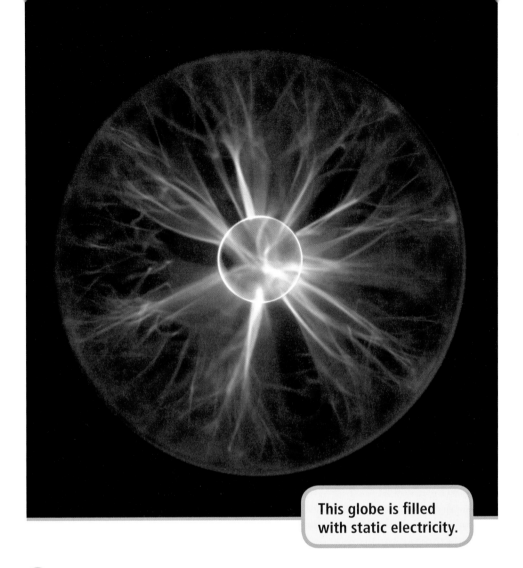

This globe is filled with static electricity.

Summary

Static electricity develops when an electrical charge builds up in an object. Current electricity travels through wires from one place to another. Magnetic items attract objects with iron in them. Electromagnets work only when electricity flows through the magnet. Generators make electricity using magnets.

Glossary

circuit (SER•kuht) A path that electricity follows (4, 15)

electricity (ee•lek•TRIS•ih•tee) A form of energy produced by moving electrons (2, 3, 4, 5, 6, 7, 12, 13, 14, 15)

magnetic (mag•NET•ik) Attracting objects that have iron in them (8, 11, 15)